The Science of Radioactive Pollution: Insights for the General Reader

Nasier

The Science of Radioactive Pollution: Insights for the General Reader

Copyright © 2023 by Nasier

The first edition was published in 2023

ISBN:
Published by:
Sunshine
1663 Liberty Drive
Hyderabad, IN 47403
www.Sunshinepublishers.com

This book is self-published using on-demand printing and publishing, which allows it to be printed and distributed globally

TABLE OF CONTENT

Chapter 3: Health Impacts of Radioactive Pollution   25

Chapter 4: Radioactive Pollution and Environmental Consequences   33

Chapter 5: Monitoring and Mitigation Strategies 42

Chapter 6: Public Awareness and Education 53

Chapter 7: Case Studies and Success Stories   61

Chapter 8: Future Perspectives and Concluding Remarks   69

Chapter 1: Introduction to Radioactive Pollution

Understanding Radioactivity

Radioactivity is a fascinating and complex phenomenon that has garnered significant attention in recent times due to its implications for pollution biology. In this subchapter, we will delve into the concept of radioactivity, explaining its origin, types, and the potential dangers it poses to the environment and living organisms. This understanding is crucial for anyone interested in pollution biology and its impact on our world.

At its core, radioactivity refers to the spontaneous emission of radiation from the unstable atomic nuclei of certain elements. These elements, known as radioisotopes, have an excess of either protons or neutrons, making them unstable and prone to decay. Through this decay process, these isotopes release energy in the form of radiation, which can take the form of alpha particles, beta particles, or gamma rays.

Alpha particles consist of two protons and two neutrons and have a positive charge. They are relatively large and can be stopped by a sheet of paper or the outer layer of human skin. Beta particles, on the other hand, are high-energy electrons or positrons released during the decay of a radioisotope. They have a negative charge and can penetrate deeper into materials than alpha particles, requiring a thicker barrier such as aluminum to stop them.

Gamma rays are the most energetic form of radiation, often emitted alongside alpha and beta particles. They are similar to X-rays but with higher energy and can easily penetrate through several centimeters of materials, including human tissue.

The impact of radioactivity on pollution biology cannot be overstated. Radioactive pollutants can enter the environment through various sources such as nuclear power plants, mining activities, and improper disposal of radioactive waste. Once released, these pollutants can contaminate air, water, and soil, affecting both flora and fauna.

Living organisms exposed to radioactivity can suffer from various health effects, including DNA damage, cell mutations, and an increased risk of cancer. The extent of these effects depends on factors such as the type and intensity of radiation, duration of exposure, and the organism's sensitivity.

In conclusion, understanding radioactivity is essential for comprehending the implications of pollution biology. By grasping the basics of radioactivity, such as its types and potential dangers, individuals interested in this field can gain insights into the impact of radioactive pollution on the environment and living organisms. With further research and knowledge, we can strive towards better practices and policies to mitigate the harmful effects of radioactivity and safeguard our planet.

Sources and Types of Radioactive Pollution

Radioactive pollution is a concerning issue that affects the environment and human health. It is essential to understand the sources and types of radioactive pollution to effectively address and mitigate its impacts. This subchapter will provide a comprehensive overview of the various sources and types of radioactive pollution.

One of the primary sources of radioactive pollution is nuclear power plants. These plants generate electricity by harnessing the energy released from nuclear reactions. However, during this process, radioactive waste is produced, which can contaminate the environment if not managed properly. Another significant source is nuclear weapons testing. The detonation of nuclear bombs releases a vast amount of radioactive material into the atmosphere, soil, and water bodies, leading to long-term contamination.

Medical practices involving the use of radioactive isotopes also contribute to radioactive pollution. Diagnostic procedures such as X-rays and CT scans emit low-level radiation, which can accumulate over time. Additionally, radioactive materials used in cancer treatments can also be a source of pollution if not handled and disposed of correctly.

Mining and extraction of radioactive elements, such as uranium and radium, can result in the release of radioactive pollutants into the environment. These activities can contaminate soil, water, and air, posing risks to both human and ecological health.

Radioactive pollution can take various forms, depending on the type of radioactive material and its mode of release. One common type is airborne pollution, where radioactive particles are released into the atmosphere and can be carried over long distances by wind currents. This type of pollution poses a significant risk as it can be inhaled or ingested by humans and animals, leading to internal radiation exposure.

Waterborne pollution occurs when radioactive materials enter water bodies through industrial discharges, accidental spills, or improper disposal of radioactive waste. This type of pollution can contaminate drinking water sources, aquatic ecosystems, and agricultural lands, affecting the entire food chain.

Soil contamination is another form of radioactive pollution, typically associated with mining activities or improper disposal of radioactive waste. Contaminated soil can lead to the uptake of radioactive elements by plants, which can then be transferred to animals and humans through the food chain.

In conclusion, understanding the sources and types of radioactive pollution is crucial for addressing and mitigating its adverse effects. Nuclear power plants, weapons testing, medical practices, and mining activities are all significant contributors to radioactive pollution. Airborne, waterborne, and soil contamination are the primary forms through which radioactive pollutants spread and affect the environment and human health. By raising awareness about these sources and types of pollution, we can work towards developing sustainable solutions and minimizing the risks associated with radioactive pollution.

Historical Background of Radioactive Pollution

Radioactive pollution is a significant environmental issue that has had far-reaching consequences on both human health and the natural world. To fully understand the gravity of this problem, it is essential to delve into its historical background and trace its origins.

The discovery of radioactivity can be attributed to the pioneering work of Marie Curie and her husband Pierre Curie in the late 19th and early 20th centuries. Through their diligent research, the Curies identified and isolated radioactive elements such as polonium and radium. Their groundbreaking work earned Marie Curie two Nobel Prizes, making her the first woman to receive this prestigious accolade.

Following this discovery, the scientific community began exploring the potential applications of radioactive materials. Radioactive substances were widely used in various industries, including medicine, agriculture, and manufacturing. Unfortunately, at the time, little was known about the long-term effects of radioactivity on both humans and the environment.

One of the most devastating incidents in the history of radioactive pollution occurred in 1945 during World War II. The United States dropped atomic bombs on the Japanese cities of Hiroshima and Nagasaki, resulting in widespread destruction and the release of a massive amount of radioactive material into the atmosphere. The immediate impact was catastrophic, causing significant loss of life and leaving survivors grappling with long-term health effects.

In subsequent decades, nuclear power emerged as a viable alternative to conventional energy sources. However, the nuclear industry also brought about its fair share of radioactive pollution incidents. The Chernobyl disaster of 1986 in Ukraine and the Fukushima nuclear accident in 2011 in Japan are two of the most well-known examples. These incidents highlighted the potential dangers associated with nuclear power and raised concerns about the long-term effects of radioactive contamination.

The historical background of radioactive pollution serves as a sobering reminder of the importance of understanding and mitigating its impact. Today, scientists and policymakers strive to develop effective strategies to minimize the release of radioactive materials into the environment and to manage existing sources of pollution. By studying the past, we can learn from our mistakes and work towards a future that prioritizes the protection of both human health and the delicate ecosystems that surround us.

In conclusion, the historical background of radioactive pollution is a crucial aspect of understanding the current state of this environmental issue. By examining the origins of radioactivity and the major incidents that have shaped its history, we can better comprehend the challenges we face today. As the field of pollution biology continues to evolve, it is essential for everyone to be aware of the historical context to drive meaningful change and safeguard our planet for future generations.

Importance of Studying Radioactive Pollution Biology

Title: Importance of Studying Radioactive Pollution Biology

Introduction:

Radioactive pollution is a significant environmental concern that poses risks to both human health and ecosystems. Understanding the impact of radioactive pollutants on living organisms is crucial in developing effective mitigation and remediation strategies. This subchapter explores the importance of studying radioactive pollution biology and its implications for pollution biology as a whole.

1. Understanding the Effects on Organisms:

Studying radioactive pollution biology helps us comprehend the direct and indirect effects of radiation on various organisms. From microscopic bacteria to complex ecosystems, understanding how radiation affects different organisms is critical in evaluating the extent of damage caused. By studying the impacts on plants, animals, and humans, researchers can develop comprehensive models to assess the risks and devise appropriate countermeasures.

2. Evaluating Ecological Consequences:

Radioactive pollution has the potential to disrupt entire ecosystems, leading to long-lasting ecological consequences. By studying the biology of radioactive pollution, scientists can assess the effects on biodiversity, food webs, and the overall health of ecosystems. Such knowledge is essential for devising conservation strategies and implementing remediation measures to restore the balance in affected areas.

3. Assessing Human Health Risks:

Radioactive pollutants can have detrimental effects on human health, ranging from acute radiation sickness to long-term genetic mutations and cancer development. Understanding the biological mechanisms through which radiation affects human cells and tissues is crucial in developing effective protection measures for individuals working in radiation-prone environments. Additionally, studying the biology of radioactive pollution enables scientists to assess the potential risks to human populations living near contaminated areas.

4. Enhancing Pollution Biology Knowledge:

Studying radioactive pollution biology contributes to the broader field of pollution biology by providing insights into the mechanisms of environmental contamination. Radiation differs from other pollutants in its unique properties and the way it interacts with living organisms. By exploring the biology of radioactive pollution, scientists can gain a deeper understanding of pollutant behavior, bioaccumulation, and ecological interactions, which can inform research and mitigation strategies for other types of pollution.

Conclusion:

The study of radioactive pollution biology is vital for understanding the impacts of radiation on living organisms, evaluating ecological consequences, assessing human health risks, and enhancing pollution biology knowledge. By delving into this field, scientists can develop effective strategies to mitigate the risks associated with radioactive pollution and work towards a healthier and safer environment for all.

Chapter 2: Basics of Radioactive Pollution Biology

The Role of Radiation in Biological Systems

Radiation is a fascinating and often misunderstood phenomenon that plays a significant role in biological systems. In this subchapter, we will explore the various ways in which radiation influences and impacts living organisms. Whether you are a curious individual or a student of pollution biology, understanding the role of radiation in biological systems is essential for comprehending the science behind radioactive pollution.

Radiation, in its various forms, can have both positive and negative effects on living organisms. On one hand, radiation is instrumental in many natural processes, such as photosynthesis, which is crucial for the growth of plants. It also plays a role in the Earth's climate by influencing the planet's energy balance. However, when radiation exceeds natural levels, it can have detrimental effects on biological systems.

One of the primary concerns related to radiation is its potential to cause genetic mutations. High-energy radiation, such as ionizing radiation, can damage the DNA molecules within cells. These mutations can lead to various health issues, including cancer and birth defects. Understanding the mechanisms behind radiation-induced mutations is crucial for developing strategies to mitigate the risks associated with radioactive pollution.

Furthermore, radiation can also affect the functioning of cells and tissues. The energy from radiation can disrupt cellular processes, leading to cell death or malfunction. This can impact the overall health and wellbeing of living organisms, including humans. Radiation exposure is a significant concern for those working in industries involving radioactive materials and for individuals living near nuclear power plants or areas with high natural background radiation.

It is important to note that not all radiation is harmful. In fact, low levels of radiation exposure can stimulate adaptive responses within biological systems, potentially enhancing their resistance to certain diseases. This phenomenon, known as radiation hormesis, has been the subject of extensive research and remains a topic of scientific debate. Understanding the complexities of radiation hormesis is vital for developing a comprehensive understanding of radiation's role in biological systems.

In conclusion, radiation plays a significant role in biological systems, both positive and negative. While it is essential for various natural processes, excessive radiation exposure can have detrimental effects on living organisms. By studying the mechanisms behind radiation-induced mutations, cellular disruptions, and radiation hormesis, we can gain valuable insights into the science of radioactive pollution. This knowledge is crucial for developing strategies to minimize the risks associated with radiation exposure and for fostering a healthier and safer environment for all living organisms.

Effects of Radioactive Pollution on Ecosystems

Radioactive pollution poses a significant threat to ecosystems around the world. As our understanding of radiation and its impact on living organisms continues to grow, it becomes increasingly important for everyone, especially those interested in pollution biology, to grasp the effects this pollution has on our natural world.

One of the most immediate and visible consequences of radioactive pollution on ecosystems is the disruption of the food chain. Radiation can accumulate in plants, which are then consumed by herbivores. As these herbivores are consumed by carnivores, the radiation is further concentrated, resulting in higher levels of contamination at the top of the food chain. This not only affects the health and survival of individual organisms, but it also has cascading effects throughout the entire ecosystem.

Another significant impact of radioactive pollution is the alteration of reproductive cycles and genetic mutations in plants and animals. Exposure to radiation can lead to genetic damage, including DNA mutations and chromosomal abnormalities. These genetic changes can result in reduced fertility, birth defects, and even the extinction of certain species. The long-term consequences of these genetic alterations are still not fully understood, making the study of radioactive pollution an ongoing and crucial field of research.

Furthermore, radioactive pollution can also affect the abundance and diversity of species within an ecosystem. Some organisms are more sensitive to radiation than others, leading to imbalances in population sizes and shifts in species composition. This disruption of biodiversity can have far-reaching ecological consequences, as each species plays a unique role in maintaining the overall health and stability of an ecosystem.

It is essential for everyone to recognize the potential long-term effects of radioactive pollution on ecosystems. By understanding these impacts, we can develop strategies to mitigate and manage the contamination, as well as work towards preventing further pollution. This requires interdisciplinary collaboration between scientists, policymakers, and the general public to raise awareness, implement stricter regulations, and invest in clean energy alternatives.

In conclusion, radioactive pollution has profound effects on ecosystems, ranging from disruptions in the food chain and genetic mutations to alterations in population sizes and biodiversity. Recognizing the importance of these effects is crucial for anyone interested in pollution biology. By understanding the consequences of radioactive pollution, we can take proactive steps to protect and restore our fragile ecosystems for the benefit of present and future generations.

Bioaccumulation and Biomagnification of Radioactive Substances

In the realm of pollution biology, one of the most concerning aspects of radioactive pollution is the phenomenon of bioaccumulation and biomagnification. This subchapter aims to shed light on these processes, helping the general reader understand the intricate relationship between radioactive substances and the living organisms that inhabit our planet.

Bioaccumulation refers to the gradual buildup of radioactive materials in an organism's body over time. When radioactive substances are released into the environment, they can be taken up by plants and animals through various pathways. For instance, plants may absorb radioactive particles from contaminated soil or water, while animals can ingest these substances through their diet or by breathing contaminated air. Once inside an organism, these radioactive materials can persist and accumulate in different tissues or organs, posing long-term health risks.

Biomagnification, on the other hand, is the process by which the concentration of radioactive substances increases at higher levels of the food chain. When organisms at the lower trophic levels consume plants or other organisms contaminated with radioactive substances, they accumulate these materials in their bodies. As predators consume these contaminated organisms, they inherit the accumulated radioactive substances, resulting in an even higher concentration. This amplification of radioactive substances as they move up the food chain can have severe consequences for top predators, including humans.

The implications of bioaccumulation and biomagnification are far-reaching. Humans, as the top predators in many ecosystems, are particularly vulnerable to the effects of biomagnification. Consuming contaminated fish, for example, can lead to the accumulation of radioactive substances in our bodies. Over time, this can lead to serious health issues, including an increased risk of cancer and other radiation-related illnesses.

Understanding the processes of bioaccumulation and biomagnification is crucial for designing effective strategies to mitigate the impact of radioactive pollution. By identifying the sources of pollution, assessing the pathways through which radioactive substances enter ecosystems, and monitoring the concentrations of these substances in organisms at different trophic levels, scientists can work towards minimizing the risks associated with exposure to radioactive materials.

In conclusion, bioaccumulation and biomagnification of radioactive substances pose significant challenges in the field of pollution biology. By comprehending these processes, we can make informed decisions and take necessary actions to protect ourselves and the environment from the harmful effects of radioactive pollution.

Factors Influencing the Impact of Radioactive Pollution on Living Organisms

Radioactive pollution, resulting from the release of radioactive materials into the environment, poses a significant threat to all living organisms. Understanding the factors that influence the impact of radioactive pollution is crucial to developing effective strategies for minimizing its harmful effects. In this subchapter, we will explore these factors and their implications for pollution biology.

One of the primary factors influencing the impact of radioactive pollution is the type of radioactive material involved. Different radioactive isotopes have varying levels of radioactivity and half-lives, which determine their persistence in the environment. Some isotopes, such as iodine-131 and cesium-137, tend to accumulate in specific organs or tissues, leading to localized damage. Others, like uranium-238, have a long half-life, resulting in prolonged exposure and potential health risks.

The concentration and duration of exposure to radioactive materials also play a crucial role in determining the impact on living organisms. Higher concentrations of radioactive pollutants can cause immediate and severe damage, while long-term exposure to lower levels can lead to chronic health problems. Additionally, the duration of exposure influences the cumulative effect of radiation on organisms, affecting their ability to repair DNA damage and recover from radiation-related injuries.

The sensitivity of different organisms to radiation is another vital factor. Certain species, such as bacteria and fungi, have mechanisms to repair DNA damage more efficiently, making them more resistant to radiation. On the other hand, highly specialized and complex organisms like humans are more susceptible to radiation-induced health problems. Factors such as age, gender, and pre-existing health conditions also influence an organism's vulnerability to radiation.

Environmental conditions can further modify the impact of radioactive pollution. Factors like temperature, humidity, and the presence of other pollutants can interact with radiation, altering its effects on living organisms. For example, high temperatures can increase the volatility of radioactive isotopes, leading to their dispersion over a wider area.

In conclusion, understanding the factors that influence the impact of radioactive pollution on living organisms is crucial for pollution biology. The type and concentration of radioactive material, duration of exposure, sensitivity of organisms, and environmental conditions all play a significant role. By comprehending these factors, scientists can develop targeted strategies to mitigate the harmful effects of radioactive pollution and protect the health of both humans and the environment.

Chapter 3: Health Impacts of Radioactive Pollution

Radiation and Human Health

Radiation is a topic that often raises concerns among people due to its potential impact on human health. In this subchapter, we will explore the science behind radiation and its effects on the human body, shedding light on the complex relationship between radiation and human health.

Radiation can be defined as the emission of energy through waves or particles. While radiation exists naturally in our environment, human activities have significantly increased exposure levels over the past century. This increase, coupled with the potential dangers associated with radiation, has sparked interest in understanding its effects on human health.

Radiation can be categorized into two types: ionizing and non-ionizing radiation. Ionizing radiation, such as X-rays and gamma rays, have enough energy to remove tightly bound electrons from atoms, leading to the formation of charged particles called ions. This type of radiation is known to have harmful effects on living organisms, including humans, as it can damage cells and DNA.

The effects of ionizing radiation on human health can vary depending on the dose and duration of exposure. Acute high-dose exposure can lead to radiation sickness, characterized by symptoms like nausea, vomiting, and fatigue. Long-term exposure to lower levels of ionizing radiation is associated with an increased risk of developing certain types of cancer, such as leukemia and thyroid cancer.

On the other hand, non-ionizing radiation, such as radio waves and microwaves, has lower energy levels and does not have enough energy to remove electrons from atoms. While non-ionizing radiation is generally considered safe, prolonged exposure to high levels of certain non-ionizing radiation, such as ultraviolet (UV) radiation from the sun, can increase the risk of skin cancer and other health issues.

To protect ourselves from the potential harm of radiation, it is crucial to understand and follow safety guidelines. For instance, individuals working with radiation in various industries undergo strict safety protocols and wear protective gear to minimize exposure. Additionally, public health initiatives emphasize the importance of limiting exposure to harmful radiation sources, such as excessive X-ray scans or prolonged sun exposure.

In conclusion, radiation and its effects on human health are complex and multifaceted. While some forms of radiation can be harmful, others are relatively safe. Understanding the different types of radiation and their potential risks can help us make informed decisions and take necessary precautions to protect ourselves and our environment. By staying informed and following safety guidelines, we can mitigate the potential health impacts of radiation exposure in our daily lives.

Types of Radiation-Induced Diseases

Radiation-induced diseases are a significant concern in today's world, where radioactive pollution is a growing issue. This subchapter will explore the various types of diseases that can be caused by exposure to radiation. It is essential for everyone, especially those interested in pollution biology, to understand the potential harm caused by radiation and its long-term effects on human health.

1. Cancer: Perhaps the most well-known radiation-induced disease is cancer. Exposure to ionizing radiation can damage DNA in our cells, leading to the development of cancerous cells. Different types of cancers, such as leukemia, thyroid cancer, and lung cancer, have been linked to radiation exposure. The risk of developing cancer increases with higher levels and prolonged exposure to radiation.

2. Genetic disorders: Another category of radiation-induced diseases includes genetic disorders. Radiation can cause mutations in the DNA of reproductive cells, leading to hereditary disorders in future generations. These disorders can range from physical abnormalities to mental impairments and can have a significant impact on the affected individuals and their families.

3. Acute Radiation Syndrome: Acute Radiation Syndrome (ARS) occurs when a person receives a high dose of radiation over a short period. Symptoms can include nausea, vomiting, diarrhea, and in severe cases, organ failure and death. ARS is a result of the direct damage to cells and tissues caused by high levels of radiation exposure.

4. Non-cancerous diseases: Radiation exposure can also lead to non-cancerous diseases such as cataracts, cardiovascular problems, and skin disorders. These diseases are often a result of damage to specific organs or tissues due to radiation.

5. Long-term effects on the immune system: Prolonged exposure to radiation can weaken the immune system, making individuals more susceptible to infections and diseases. This can result in higher mortality rates and a reduced quality of life for those affected.

Understanding the different types of radiation-induced diseases is crucial in developing effective strategies for prevention, mitigation, and treatment. It is imperative for the general reader, especially those interested in pollution biology, to be aware of the potential risks associated with radioactive pollution. By educating ourselves and implementing appropriate safety measures, we can work towards minimizing the impact of radiation-induced diseases on human health and the environment.

In conclusion, radiation-induced diseases encompass a wide range of health issues, including cancer, genetic disorders, acute radiation syndrome, non-cancerous diseases, and long-term effects on the immune system. By understanding these diseases, we can take necessary precautions and work towards minimizing the impact of radioactive pollution on our health and well-being.

Long-term Effects of Radioactive Exposure

Radioactive pollution is an issue that affects not only our environment but also our health. The long-term effects of radioactive exposure can be devastating and far-reaching. In this subchapter, we will explore the various ways in which radiation can impact our bodies and the environment, shedding light on the science behind these effects.

One of the most well-known long-term effects of radioactive exposure is an increased risk of cancer. Ionizing radiation, emitted by radioactive materials, has the ability to damage our DNA, leading to the development of cancerous cells. Studies have shown a clear correlation between exposure to radiation and the incidence of certain types of cancer, such as leukemia and thyroid cancer. These findings highlight the importance of understanding the risks associated with radioactive pollution and implementing measures to minimize exposure.

Aside from cancer, exposure to radiation can also have other health effects. It can weaken the immune system, making individuals more susceptible to infections and diseases. Additionally, radiation can affect the reproductive system, leading to fertility issues and an increased risk of birth defects in offspring. These long-term effects highlight the need for stringent regulations and monitoring of radioactive materials to safeguard public health.

Furthermore, the environment is also greatly impacted by radioactive pollution. Plants and animals living in contaminated areas can suffer from genetic mutations, reduced reproductive capabilities, and even population decline. The accumulation of radioactive materials in the soil and water can persist for years, affecting the entire ecosystem. Understanding the long-term effects of radioactive exposure on the environment is crucial for conservation efforts and the preservation of biodiversity.

As we delve into the long-term effects of radioactive exposure, it is important to note that there are ways to mitigate these risks. Implementing proper safety measures and regulations in industries that handle radioactive materials is essential. Educating the public about the potential dangers of radiation and promoting responsible disposal of radioactive waste can also help minimize the long-term effects of exposure.

In conclusion, the long-term effects of radioactive exposure can have significant implications for both human health and the environment. Understanding the science behind these effects is crucial for pollution biologists and anyone concerned with the well-being of our planet. By raising awareness and implementing effective strategies, we can work towards minimizing the risks associated with radioactive pollution and ensuring a safer, healthier future for all.

Risk Assessment and Management of Radioactive Pollution

Radioactive pollution is a significant environmental concern that poses serious risks to both human health and the ecosystem. Understanding how to assess and manage these risks is essential for protecting ourselves and our planet. This subchapter explores the science behind risk assessment and management of radioactive pollution, providing insights for the general reader, particularly those interested in pollution biology.

Risk assessment is the process of evaluating the potential dangers associated with exposure to radioactive materials and determining the likelihood of adverse effects. It involves examining various factors such as the type of radioactive material, its concentration in the environment, the duration and route of exposure, and the vulnerability of the exposed population. By quantifying these factors, scientists can estimate the risks and develop appropriate strategies for managing them.

The first step in risk assessment is to measure the levels of radioactivity in affected areas. This is done through environmental monitoring, which involves using specialized equipment to detect and quantify radioactive substances. These measurements help scientists determine the extent of contamination and assess the potential risks to humans, plants, and animals.

Once the risks are identified, effective management strategies can be implemented. Risk management aims to minimize exposure to radioactive materials and prevent or mitigate their adverse effects. This can be achieved through various methods, such as containment and remediation of contaminated sites, implementing safety measures in industries dealing with radioactive materials, and establishing regulations and guidelines for their safe handling and disposal.

It is crucial to involve multiple stakeholders, including scientists, government agencies, industries, and the public, in the risk assessment and management process. Open communication, education, and public awareness campaigns play a vital role in ensuring that individuals understand the risks associated with radioactive pollution and take appropriate measures to protect themselves and the environment.

In conclusion, understanding the risks associated with radioactive pollution is essential for effective management and prevention of its adverse effects. By employing rigorous risk assessment techniques and implementing appropriate management strategies, we can minimize exposure to radioactive materials and safeguard the health of ecosystems and human populations. This subchapter serves as a comprehensive guide for a general audience interested in pollution biology, providing valuable insights into the science behind risk assessment and management of radioactive pollution.

Chapter 4: Radioactive Pollution and Environmental Consequences

Radioactive Pollution and Climate Change

Radioactive pollution is a significant concern for our environment and has far-reaching implications for both human and ecological health. In recent years, the impact of radioactive pollution on climate change has gained attention and raised alarming questions about the long-term consequences of our actions.

Radioactive pollution refers to the release of radioactive substances into the environment, either through natural processes or as a result of human activities such as nuclear power generation, mining, or nuclear accidents. These substances can contaminate air, water, and soil, posing serious risks to living organisms.

The connection between radioactive pollution and climate change lies in the release of greenhouse gases during the process of nuclear power generation. The burning of fossil fuels for electricity production is a major contributor to greenhouse gas emissions, which results in global warming and climate change. However, some argue that nuclear power, touted as a cleaner alternative, also contributes to climate change due to the release of carbon dioxide during the mining, processing, and transportation of uranium.

Moreover, the potential for nuclear accidents, like the devastating Chernobyl and Fukushima incidents, poses a significant threat to the environment and climate. These accidents release large amounts of radioactive materials into the atmosphere, leading to immediate and long-term consequences. The radioactive substances released during such accidents can contaminate ecosystems, affecting the biodiversity and overall balance of the environment.

The impact of radioactive pollution on climate change extends beyond the immediate aftermath of accidents. It can also have long-term effects on the climate system, altering weather patterns and contributing to the destabilization of ecosystems. The release of radioactive substances can lead to the destruction of vegetation and the disruption of natural processes, ultimately affecting the balance of carbon dioxide and oxygen levels in the atmosphere.

To mitigate the impact of radioactive pollution on climate change, it is crucial to adopt sustainable practices and invest in renewable energy sources. By reducing our dependence on nuclear power and transitioning to clean energy alternatives like solar and wind power, we can minimize the release of greenhouse gases and promote a healthier environment.

In conclusion, radioactive pollution and climate change are interconnected issues that demand our attention. The consequences of radioactive pollution on climate change are significant, and it is crucial for us to understand the intricate link between the two. By raising awareness about the potential risks and promoting sustainable practices, we can work towards a safer and healthier environment for all.

Impact of Radioactive Pollution on Biodiversity

Subchapter: Impact of Radioactive Pollution on Biodiversity

Introduction:
Radioactive pollution is a significant environmental issue that poses a threat to all forms of life on Earth. In this subchapter, we will explore the impact of radioactive pollution on biodiversity, shedding light on its consequences for various species and ecosystems. By understanding the far-reaching effects of this type of pollution, we can better comprehend the urgency of mitigating its harmful outcomes.

1.   Radioactive   Pollution   and   Biodiversity: Radioactive pollutants, such as radioactive isotopes of elements like uranium, plutonium, and cesium, can be released into the environment through accidents at nuclear power plants, improper disposal of nuclear waste, or nuclear weapons testing. These pollutants can contaminate air, water, and soil, thereby affecting biodiversity in numerous ways.

2.   Direct   Effects   on   Species: High levels of radiation exposure can cause acute health problems in both plants and animals. For instance, wildlife living in contaminated areas may experience genetic mutations, birth defects, and an increased risk of cancer. Birds, mammals, and amphibians are particularly vulnerable, as their reproductive cycles and genetic diversity can be severely affected.

3.   Indirect   Effects   on   Ecosystems: Radioactive pollution can disrupt entire ecosystems, leading to a cascade of negative consequences. For example, radiation can inhibit the growth of plants, reducing food sources and habitat availability for other organisms. This disruption can then impact the populations of herbivores, predators, and decomposers, ultimately affecting the balance of the entire ecosystem.

4.   Bioaccumulation   and   Biomagnification: Radioactive pollutants have the potential to accumulate in the tissues of living organisms. This process, known as bioaccumulation, occurs as radioactive substances are absorbed by plants and animals and stored in their bodies. Additionally, the phenomenon of biomagnification amplifies the concentration of radioactive pollutants as they move up the food chain. This means that top predators, such as apex predators or humans, may be exposed to the highest levels of radiation.

5.   Long-term   Consequences: The long-term impact of radioactive pollution on biodiversity is concerning. The genetic damage caused by radiation exposure can persist across generations, leading to population declines and reduced species diversity. These consequences can extend far beyond the immediate vicinity of the pollution source, affecting larger geographic areas and even global biodiversity.

Conclusion:
Understanding the impact of radioactive pollution on biodiversity is crucial for pollution biologists and the general public alike. By

recognizing the far-reaching consequences of this type of pollution, we can advocate for stricter regulations, improved waste management practices, and the development of cleaner energy alternatives. Only through collective efforts can we safeguard biodiversity and protect the delicate balance of our ecosystems.

Radioactive Pollution and Water Contamination

Water is one of the most essential resources for all life forms on Earth. It sustains not only human beings but also a diverse range of ecosystems. However, our water sources are increasingly under threat from various forms of pollution, including radioactive contamination. Radioactive pollution, caused by the release of radioactive substances into the environment, poses a significant risk to both human health and the ecological balance.

In this subchapter, we will explore the complex relationship between radioactive pollution and water contamination, shedding light on its causes, consequences, and potential solutions. While the topic may seem daunting, we aim to present the information in a manner accessible to readers of all backgrounds, ensuring a comprehensive understanding of the subject matter.

Radioactive pollution can enter water bodies through multiple pathways. One primary source is the improper disposal of radioactive waste from nuclear power plants, research facilities, or medical institutions. Another common source is the fallout from nuclear accidents or weapons testing. Once released into the environment, these radioactive substances can contaminate groundwater, rivers, lakes, and even oceans, affecting the entire water cycle.

The consequences of water contamination due to radioactive pollution are far-reaching. Exposure to radioactive materials can lead to severe health problems, including cancers, birth defects, and genetic mutations. Aquatic ecosystems, too, bear the brunt of this pollution, with devastating impacts on fish, algae, and other aquatic organisms. The delicate balance of these ecosystems is disrupted, leading to biodiversity loss and potential ecosystem collapse.

Addressing radioactive pollution and water contamination requires a multi-faceted approach. Strict regulations and proper waste management practices are essential to prevent further pollution. Additionally, exploring alternative energy sources, such as solar or wind power, can help reduce the reliance on nuclear energy and minimize the risks associated with it. Remediation techniques, such as phytoremediation or bioremediation, also hold promise in cleaning up contaminated water bodies.

Overall, understanding the science behind radioactive pollution and water contamination is crucial for everyone, especially those interested in pollution biology. By delving into this topic, we empower ourselves to make informed decisions, advocate for responsible practices, and work towards a cleaner and safer environment for future generations. Together, we can mitigate the impacts of radioactive pollution and protect our precious water resources.

Effects of Radioactive Pollution on Soil and Agriculture

Radioactive pollution is a growing concern in today's world, with its detrimental effects on the environment and human health becoming increasingly evident. One area severely impacted by this form of pollution is soil and agriculture. The consequences of radioactive pollution on these crucial aspects of our lives are far-reaching and demand our immediate attention.

When radioactive materials are released into the environment, they can contaminate the soil, leading to serious repercussions for agriculture. The primary source of this contamination is nuclear accidents, such as the Chernobyl disaster in 1986 and the Fukushima incident in 2011. These events released substantial amounts of radioactive particles into the air, which eventually settled onto the soil.

The presence of radioactive elements in the soil can cause a variety of problems for agricultural practices. One of the most significant concerns is the impact on crop growth and productivity. Radioactive pollution can inhibit the uptake of essential nutrients by plants, leading to stunted growth and decreased crop yields. Additionally, certain radioactive isotopes can replace essential minerals in the soil, making it unsuitable for plant growth.

Furthermore, radioactive pollution can affect the quality and safety of agricultural products. Plants grown in contaminated soil may accumulate radioactive isotopes, which can then be ingested by humans and animals. This can have severe health consequences, including an increased risk of cancer and genetic mutations.

The long-term effects of radioactive pollution on soil and agriculture are not limited to immediate contamination. Radioactive isotopes can persist in the environment for decades, continuing to pose a threat to agricultural lands and food production. This highlights the urgent need for comprehensive measures to mitigate and remediate radioactive pollution in affected areas.

Efforts to combat the effects of radioactive pollution on soil and agriculture focus on decontamination and monitoring strategies. Decontamination techniques, such as soil removal and replacement, can be employed to reduce the levels of radioactive materials in affected areas. Additionally, regular monitoring of soil and agricultural products is essential to ensure the safety of food consumed by the general population.

In conclusion, radioactive pollution has significant implications for soil and agriculture. The contamination of soil can lead to reduced crop yields, compromised food safety, and long-term environmental consequences. It is crucial for policymakers, scientists, and the general public to recognize the gravity of this issue and work towards effective solutions to mitigate the effects of radioactive pollution on soil and agriculture. By doing so, we can protect our food supply and safeguard the health of present and future generations.

Chapter 5: Monitoring and Mitigation Strategies

Radioactive Pollution Monitoring Techniques

In this subchapter, we will explore various techniques used for monitoring radioactive pollution. As concerns about the environmental impact of radioactive substances continue to grow, it becomes imperative to develop effective monitoring strategies to ensure the safety of our ecosystems and the well-being of all living organisms.

One commonly employed technique is aerial monitoring, which involves the use of aircraft equipped with specialized detectors to measure the radiation levels from above. This method is particularly useful in identifying areas of high contamination, such as nuclear power plants or radioactive waste storage sites. Aerial monitoring allows for quick and wide-ranging assessments, making it an essential tool in emergency situations or during large-scale radioactive incidents.

Another technique is ground-based monitoring, which involves placing stationary detectors on the ground to continuously measure radiation levels. This method provides long-term data on radiation levels in specific areas and helps identify any changes or trends over time. Ground-based monitoring allows for the identification of hotspots and helps authorities take appropriate measures to mitigate the risks posed by radioactive pollution.

In addition to aerial and ground-based monitoring, there are also techniques for monitoring radioactive pollution in water bodies. Water sampling and analysis provide crucial information about the presence and concentration of radioactive substances in rivers, lakes, and oceans. Monitoring water bodies is crucial as radioactive pollutants can spread through water sources and contaminate aquatic ecosystems, potentially affecting the entire food chain.

Furthermore, biological monitoring techniques involve the analysis of organisms living in areas affected by radioactive pollution. By studying the health and genetic makeup of these organisms, scientists can gain insights into the impacts of radiation exposure on living organisms. Biological monitoring provides valuable data on the long-term effects of radioactive pollution and helps assess the overall health of ecosystems.

Lastly, remote sensing techniques using satellites and other advanced technologies offer a unique perspective on monitoring radioactive pollution. These techniques allow for the collection of data from large areas, providing a comprehensive view of the distribution and extent of radioactive contamination. Remote sensing techniques are particularly useful in monitoring areas that are difficult to access or have limited ground-based monitoring infrastructure.

In conclusion, the monitoring of radioactive pollution is vital to safeguarding our environment and ensuring the well-being of all living beings. A combination of aerial, ground-based, water, biological, and remote sensing techniques provides a comprehensive approach to monitoring radioactive pollution. By employing these techniques, scientists and policymakers can make informed decisions to mitigate the risks associated with radioactive substances and protect our ecosystems for future generations.

Regulations and Guidelines for Radioactive Pollution Control

Radioactive pollution poses significant risks to the environment and human health. Therefore, strict regulations and guidelines are necessary to ensure the safe handling and disposal of radioactive materials. This subchapter aims to provide an overview of the key regulations and guidelines implemented to control radioactive pollution.

1. International Atomic Energy Agency (IAEA) Standards: The IAEA plays a crucial role in setting international standards for the safe use, transportation, and disposal of radioactive materials. These standards help countries establish their national regulations and guidelines for radioactive pollution control.

2. National Regulatory Authorities: Each country has its own regulatory authority responsible for overseeing the safe use of radioactive materials. These authorities establish regulations and guidelines that cover areas such as licensing, radiation protection, waste management, and emergency preparedness.

3. Radiation Protection Standards: To protect human health and the environment, regulatory authorities set limits on the allowable exposure to radiation. These limits are based on extensive research and take into account various factors such as the type of radiation, duration of exposure, and potential health risks.

4. Waste Management: Proper management of radioactive waste is vital to prevent contamination of the environment. Regulations and

guidelines require the classification, segregation, and safe storage of radioactive waste. Additionally, disposal methods, such as deep geological repositories or encapsulation in durable materials, must be in compliance with established standards.

5. Emergency Preparedness: Contingency plans and emergency response procedures are essential in case of accidents or incidents involving radioactive materials. Regulatory authorities mandate that facilities handling radioactive materials have robust emergency preparedness plans in place to minimize the impact of any potential release of radioactive substances.

6. Monitoring and Reporting: Regular monitoring of radiation levels in the environment is necessary to identify potential sources of radioactive pollution. Regulatory authorities require facilities to conduct periodic monitoring and submit reports detailing their findings. This information allows for prompt action to be taken if any deviations from the established standards are detected.

7. Public Awareness and Education: Regulations also emphasize the importance of public awareness and education regarding radioactive pollution. It is crucial for individuals to understand the risks associated with exposure to radiation and the measures in place to protect them. By fostering public awareness, regulatory authorities aim to encourage responsible behavior and ensure cooperation with necessary safety measures.

In conclusion, regulations and guidelines for radioactive pollution control are essential to safeguard the environment and human health.

These measures are implemented at both international and national levels to ensure the safe handling, transportation, and disposal of radioactive materials. By adhering to these regulations and guidelines, we can effectively mitigate the risks posed by radioactive pollution and protect our ecosystems for future generations.

Remediation and Cleanup of Radioactive Contaminated Sites

Radioactive pollution is a global concern, as it poses significant risks to human health and the environment. The contamination of sites with radioactive substances requires prompt remediation and cleanup efforts to minimize these risks. In this subchapter, we will explore the various methods and technologies used in the remediation and cleanup of radioactive contaminated sites.

One commonly used method is containment, which involves isolating the contaminated area to prevent the spread of radioactive materials. This can be achieved through the construction of physical barriers, such as concrete walls or underground enclosures. Containment is an effective short-term solution, but it does not eliminate the source of contamination.

Another approach is the physical removal of contaminated materials. This method involves excavating and removing the contaminated soil or debris from the site. The materials are then transported to a secure facility for proper disposal or treatment. Physical removal is often combined with other techniques to ensure thorough cleanup.

Chemical treatment is another method used in the remediation process. It involves the use of chemicals to dissolve or neutralize radioactive substances, making them less hazardous. For instance, certain chemicals can be used to bind with radioactive elements, preventing their migration and reducing their mobility. Chemical treatment can be applied to both soil and water contamination.

Bioremediation is a promising technique that utilizes living organisms, such as bacteria and plants, to degrade or absorb radioactive substances. Certain microorganisms have the ability to break down radioactive contaminants into non-toxic forms. Additionally, some plant species can accumulate radioactive materials in their tissues, a process known as phytoextraction.

In situ immobilization is another method used to remediate radioactive contaminated sites. This technique involves the addition of substances that bind with radioactive elements, reducing their mobility. These substances can be natural minerals or specially designed materials. In situ immobilization is often used in conjunction with other methods to enhance their effectiveness.

Overall, the remediation and cleanup of radioactive contaminated sites require a multidisciplinary approach. Scientists, engineers, and environmental experts work together to develop and implement effective strategies. It is crucial to consider factors such as site-specific conditions, the type and extent of contamination, and the potential long-term effects. Through the application of these methods and technologies, we can mitigate the risks associated with radioactive pollution and restore the affected areas to a safe and healthy state.

This subchapter provides valuable insights into the various techniques used in the remediation and cleanup of radioactive contaminated sites. It is essential reading for anyone interested in pollution biology and the preservation of our environment. By understanding these methods, we can contribute to the ongoing efforts to combat radioactive pollution and protect the well-being of present and future generations.

50

Future Directions in Radioactive Pollution Management

As we continue to expand our knowledge and understanding of radioactive pollution, it becomes crucial to explore future directions in its management. The adverse effects of radioactive pollution on the environment and human health have long been a concern in the field of pollution biology. To address these challenges, scientists and researchers are striving to develop innovative solutions that can mitigate the impact of radioactive pollution and ensure a safer and healthier world for future generations.

One of the key areas of focus in future radioactive pollution management is the development of advanced monitoring and detection techniques. These techniques will enable us to identify and quantify radioactive contaminants more accurately and promptly. By employing cutting-edge technologies such as remote sensing, drones, and artificial intelligence, we can enhance our ability to detect and monitor radioactive pollutants in real-time. This will allow for more efficient and targeted remediation efforts.

Another promising direction is the development of novel decontamination methods. Traditional decontamination techniques can be time-consuming, costly, and sometimes ineffective. However, researchers are exploring alternative approaches such as phytoremediation, which uses plants to remove radioactive substances from soil and water. Additionally, nanotechnology offers possibilities for the development of efficient and selective adsorbents that can capture and remove radioactive materials from the environment.

In the future, it will be essential to focus on the management of radioactive waste. As nuclear energy continues to be a significant part of our energy mix, the proper disposal of radioactive waste becomes paramount. Scientists are researching new methods for safe and long-term storage of radioactive waste, including deep geological repositories and advanced waste treatment technologies. By implementing robust waste management strategies, we can minimize the risk of radioactive contamination and ensure the protection of the environment and human health.

Collaboration and international cooperation are also crucial for effective radioactive pollution management. The global nature of radioactive pollutants requires a unified approach, with countries working together to share knowledge, resources, and best practices. International agreements and organizations, such as the International Atomic Energy Agency (IAEA), play a vital role in coordinating efforts and establishing common guidelines for radioactive pollution management.

In conclusion, the future of radioactive pollution management holds great promise. Through advancements in monitoring and detection techniques, innovative decontamination methods, proper waste management, and international collaboration, we can effectively address the challenges posed by radioactive pollution. By striving towards these future directions, we can ensure a safer and healthier environment for all and pave the way for a sustainable future.

Chapter 6: Public Awareness and Education

Communicating the Risks of Radioactive Pollution

In today's modern world, radioactive pollution has become a significant concern that affects not only the environment but also the health and well-being of all living organisms. It is essential for everyone, especially those interested in Pollution Biology, to understand the risks associated with this form of pollution. This subchapter aims to educate and inform the general reader about the science behind radioactive pollution and its potential consequences.

Radioactive pollution occurs when radioactive materials are released into the environment, either through accidents at nuclear power plants, improper disposal of nuclear waste, or nuclear weapons testing. These radioactive materials emit harmful ionizing radiation, which can have severe impacts on living organisms and ecosystems.

One of the key challenges in communicating the risks of radioactive pollution lies in the complex nature of radiation and its effects. This subchapter seeks to simplify and explain these concepts in a way that is accessible to everyone. It will delve into the different types of radiation, such as alpha, beta, and gamma radiation, and their varying levels of penetration and potential harm to living organisms.

Furthermore, this subchapter will highlight the potential health effects of exposure to radioactive pollution, including an increased risk of cancer, genetic mutations, and other long-term health issues. It will emphasize that even low levels of radiation exposure can have cumulative effects over time.

To effectively communicate these risks, it is crucial to provide practical information on how individuals can protect themselves and minimize their exposure to radioactive pollution. This subchapter will offer guidance on simple measures such as staying informed about local radiation levels, understanding the risks associated with specific activities (such as nuclear power generation or medical procedures involving radiation), and following safety protocols in areas where radioactive materials are present.

Lastly, this subchapter will stress the importance of responsible decision-making and government regulations in managing radioactive pollution. It will address the need for transparent information, public involvement in decision-making processes, and the role of scientists and experts in monitoring and mitigating the risks associated with this form of pollution.

By providing a clear and accessible understanding of the risks involved, this subchapter aims to empower individuals to make informed choices and contribute to the collective efforts in reducing radioactive pollution. It is crucial for everyone, regardless of their background or expertise, to grasp the significance of this issue and work towards a healthier and safer environment for present and future generations.

Public Perception and Misconceptions about Radioactive Pollution

Radioactive pollution is a topic that often elicits strong emotions and concerns among the general public. However, it is important to approach this issue with a clear understanding of the science behind it, in order to dispel common misconceptions and foster informed discussions. This subchapter aims to address the public perception and misconceptions surrounding radioactive pollution, providing insights for readers from all walks of life, particularly those interested in pollution biology.

One of the most prevalent misconceptions is that any level of radiation exposure is extremely dangerous and harmful. While it is true that high levels of radiation can have severe health effects, it is important to note that exposure to low levels of radiation is a part of daily life. Our environment is naturally radioactive, and we are exposed to radiation from various sources such as the sun and the Earth's crust. Understanding the concept of radiation dose and the difference between natural and man-made radiation is crucial in comprehending the actual risks associated with radioactive pollution.

Another misconception is that all radioactive substances are hazardous and should be avoided at all costs. While it is true that certain radioactive elements, such as uranium and plutonium, can pose significant risks, not all radioactive substances are equally dangerous. Some radioactive isotopes have short half-lives and decay quickly, while others have long half-lives and remain radioactive for thousands of years. It is important to differentiate between different types of radioactive materials and their potential impact on human health.

Furthermore, public perception often associates all nuclear activities, such as power generation and medical applications, with radioactive pollution. While accidents like Chernobyl and Fukushima have had devastating consequences, it is important to recognize that these incidents are rare and do not represent the norm. Nuclear technology, when properly managed and regulated, can provide numerous benefits, including clean energy generation and medical diagnostics and treatments.

In conclusion, public perception and misconceptions about radioactive pollution are widespread, and it is crucial to address them with accurate scientific information. By understanding the concepts of radiation dose, different types of radioactive materials, and the benefits and risks associated with nuclear technology, individuals can make informed decisions and contribute to meaningful discussions about radioactive pollution. This subchapter aims to provide insights for readers interested in pollution biology, as well as for a general audience seeking a deeper understanding of this complex issue.

Role of Education in Addressing Radioactive Pollution

Introduction:
Education plays a vital role in addressing the issue of radioactive pollution, as it enables individuals to understand the complexities and consequences of this environmental problem. This subchapter explores the significance of education in raising awareness, promoting responsible behavior, and fostering sustainable solutions to mitigate radioactive pollution. By disseminating knowledge and empowering individuals, education serves as a powerful tool in tackling this pressing issue.

Raising   Awareness:
Education is essential in increasing public awareness about the sources, effects, and risks associated with radioactive pollution. By educating the general public, policymakers, and stakeholders about the dangers of radiation exposure, people become more conscious of their actions and the potential consequences. This knowledge helps individuals make informed choices and take necessary precautions to minimize their exposure to radioactive materials.

Promoting   Responsible   Behavior:
Education about radioactive pollution encourages responsible behavior in individuals and communities. Through education, people learn about proper waste management techniques, radiation protection measures, and the importance of adhering to safety protocols. By promoting responsible behavior, education helps prevent the improper disposal of radioactive materials, reduces the

risk of contamination, and ensures the safety of both humans and the environment.

Fostering Sustainable Solutions:
Education fosters the development of sustainable solutions to address radioactive pollution. By providing individuals with scientific knowledge and technical skills, education equips them to contribute to research, innovation, and policy-making in this field. This enables the creation of effective strategies and technologies for the detection, monitoring, and remediation of radioactive pollution. Education also encourages interdisciplinary approaches, fostering collaboration among experts in pollution biology, environmental science, and other relevant fields.

Conclusion:
Education plays a critical role in addressing radioactive pollution by raising awareness, promoting responsible behavior, and fostering sustainable solutions. By disseminating knowledge and empowering individuals, education enables them to understand the sources and consequences of radioactive pollution. This knowledge helps individuals make informed decisions, take necessary precautions, and adopt responsible behavior to minimize their exposure to radioactive materials. Furthermore, education equips individuals with the skills and expertise to contribute to research, innovation, and policymaking, leading to the development of effective strategies for the detection, monitoring, and remediation of radioactive pollution. Therefore, education is an essential tool in combating radioactive pollution and creating a sustainable future for all.

Engaging the Public in Radioactive Pollution Prevention

In today's rapidly advancing world, the issue of radioactive pollution has become a major concern for both scientists and the general public. As awareness about the detrimental effects of radioactive materials grows, it is crucial to engage the public in preventing and mitigating the risks associated with radioactive pollution. This subchapter aims to shed light on the importance of involving every individual in the fight against this environmental hazard.

Radioactive pollution is a complex issue that requires a multidisciplinary approach. Pollution Biology, as a niche field, plays a critical role in understanding the biological impacts of radioactive materials on ecosystems and human health. By bridging the gap between scientific knowledge and the general public, Pollution Biology can effectively communicate the urgency of taking preventive measures.

Engaging the public in radioactive pollution prevention starts with education and awareness. This subchapter will provide an accessible overview of the science behind radioactive pollution, its sources, and its potential consequences. By presenting scientific information in a clear and concise manner, the book aims to empower readers from all walks of life to make informed decisions regarding their own actions and lifestyles.

Furthermore, the subchapter will explore various strategies to actively involve the public in radioactive pollution prevention. It will highlight the significance of community-based initiatives, such as citizen science projects, where individuals can actively participate in monitoring radioactive contamination in their surroundings. By encouraging public participation in data collection and analysis, we can gather valuable information on the extent and impacts of radioactive pollution.

Moreover, the subchapter will emphasize the role of advocacy and policy-making in addressing radioactive pollution. It will emphasize the need for collaboration between scientists, policymakers, and the general public to develop effective preventive measures and regulations. By engaging with local and national authorities, individuals can contribute to the formulation of policies that prioritize radioactive pollution prevention and the promotion of clean energy alternatives.

Engaging the public in radioactive pollution prevention is not just a scientific endeavor; it is a collective responsibility. This subchapter aims to inspire every reader, regardless of their background, to take an active role in reducing the risks associated with radioactive pollution. Through education, awareness, and community involvement, we can create a sustainable future where the detrimental effects of radioactive materials are minimized, and the health of our ecosystems and our own well-being are safeguarded for generations to come.

Chapter 7: Case Studies and Success Stories

Chernobyl Disaster: Lessons Learned

The Chernobyl disaster is etched in the annals of history as one of the most catastrophic nuclear accidents ever witnessed. This subchapter aims to shed light on the valuable lessons learned from this disaster, particularly in the field of pollution biology. While the incident itself brought immense devastation and loss, it also provided scientists and researchers with valuable insights into the far-reaching consequences of radioactive pollution.

One of the most significant lessons learned from Chernobyl is the undeniable impact of radioactive contamination on the environment and living organisms. The aftermath of the accident revealed the devastating effects of radiation on plants, animals, and humans alike. This event served as a wake-up call, highlighting the need for a deeper understanding of the long-term ecological consequences of radioactive pollution.

Furthermore, the Chernobyl disaster highlighted the importance of effective emergency response and preparedness. The initial response to the accident was marred by confusion and lack of knowledge regarding the proper handling of a nuclear crisis. This incident emphasized the necessity for comprehensive emergency plans and the training of professionals to handle such situations.

The disaster also underscored the importance of public awareness and education. The lack of information and understanding regarding the potential hazards of nuclear power contributed to the severity of the Chernobyl incident. In response, there was a global push to educate the general public about the risks associated with radiation and the importance of safety precautions. This incident prompted a broader conversation about the ethical responsibilities of scientists, policymakers, and the public in dealing with nuclear technology.

Additionally, the Chernobyl disaster led to advancements in the field of pollution biology. Scientists and researchers were able to study the long-term effects of radiation on various organisms, providing valuable insights into the mechanisms of radioactive contamination. These studies have since helped in developing strategies for mitigating the impact of nuclear accidents and in monitoring and assessing the ecological health of contaminated areas.

In conclusion, the Chernobyl disaster served as a turning point in the understanding and management of radioactive pollution. The lessons learned from this tragedy have led to significant advancements in pollution biology, emergency response, public awareness, and education. By drawing upon these lessons, we can strive to prevent similar incidents in the future and work towards a safer and more sustainable approach to nuclear technology.

Fukushima Daiichi Nuclear Accident: Impacts and Recovery

The Fukushima Daiichi nuclear accident, which occurred on March 11, 2011, was one of the worst nuclear disasters in history. It was triggered by a powerful earthquake and subsequent tsunami, which severely damaged the Fukushima Daiichi Nuclear Power Plant in Japan. The accident led to the release of radioactive materials into the environment, causing significant impacts on the region and raising concerns about the long-term effects of radioactive pollution.

The immediate aftermath of the Fukushima Daiichi accident was marked by widespread evacuations, as authorities scrambled to protect people from the potential health risks of radiation exposure. The accident resulted in the release of radioactive isotopes such as iodine-131 and cesium-137, which can cause various health problems, including an increased risk of cancer. The impacts of these releases were felt not only in Japan but also in neighboring countries, as airborne radioactive particles were carried by wind and deposited over large areas.

In the years following the accident, extensive monitoring and cleanup efforts were undertaken to mitigate the impacts of the radioactive pollution. These efforts included decontamination of affected areas, removal of contaminated soil, and the implementation of measures to minimize further releases of radioactive materials. These actions were essential to protect human health and restore the affected areas to a safe state.

The recovery process after the Fukushima Daiichi accident has been a complex and challenging endeavor. Apart from the physical cleanup, restoring public confidence in nuclear energy and addressing the social and psychological impacts on affected communities were crucial aspects of the recovery efforts. The accident prompted a reassessment of safety measures at nuclear power plants worldwide and led to stricter regulations and improved emergency response plans.

While significant progress has been made in the recovery process, the long-term effects of the Fukushima Daiichi accident are still being studied. Ongoing research aims to better understand the ecological and health impacts of the radioactive pollution, as well as the effectiveness of the implemented countermeasures. This knowledge will not only benefit the affected region but also contribute to global efforts in pollution biology, helping scientists and policymakers understand the potential consequences of nuclear accidents and develop strategies to mitigate their impacts.

In conclusion, the Fukushima Daiichi nuclear accident had far-reaching impacts on the environment, human health, and public perception of nuclear energy. The recovery process has involved both physical cleanup and efforts to regain public trust. Ongoing research and monitoring will continue to shed light on the long-term effects of the accident and inform future measures to prevent and mitigate the impacts of radioactive pollution.

Environmental Restoration Efforts in Radioactive Hotspots

Radioactive pollution is a significant global concern that poses long-lasting threats to both human health and the environment. In certain areas, known as radioactive hotspots, the concentration of radioactive substances exceeds safe limits, necessitating urgent environmental restoration efforts. These hotspots can occur due to various reasons, including nuclear accidents, improper disposal of radioactive waste, or historical nuclear testing.

Environmental restoration in radioactive hotspots involves a series of complex and challenging tasks aimed at reducing the impact of radioactive contamination on ecosystems and human populations. The overarching goal is to restore the affected areas to a state where they can support sustainable and healthy ecosystems, while also minimizing the risk to human health.

One of the primary methods used in environmental restoration efforts is decontamination. This process involves removing or reducing the concentration of radioactive substances from the affected area. Decontamination techniques vary depending on the specific circumstances and can include physical removal of contaminated soil, water, or vegetation, as well as chemical treatments or immobilization of radioactive elements.

In addition to decontamination, another crucial aspect of environmental restoration is the remediation of affected ecosystems. This involves the reintroduction or promotion of native plant and animal species, the restoration of habitats, and the improvement of soil quality. By restoring the ecological balance and biodiversity of the area, the ecosystem's resilience to radioactive contamination can be enhanced.

Monitoring and long-term management are also vital components of environmental restoration efforts in radioactive hotspots. Regular monitoring of radiation levels in soil, water, and air is necessary to assess the effectiveness of restoration measures and ensure that contamination levels remain within acceptable limits. Furthermore, ongoing management strategies are required to prevent further contamination and ensure the sustainability of restored ecosystems.

Environmental restoration efforts in radioactive hotspots are interdisciplinary endeavors that require collaboration between scientists, policymakers, and local communities. By sharing knowledge and resources, these efforts can be more effective in mitigating the impacts of radioactive pollution and safeguarding the environment for future generations.

In conclusion, environmental restoration in radioactive hotspots is a critical and complex undertaking that aims to reduce the impact of radioactive contamination on ecosystems and human health. Through decontamination, remediation, and long-term management strategies, these efforts strive to restore affected areas to a state of ecological health and sustainability. By understanding and supporting these restoration efforts, we can contribute to the preservation of our environment and protect ourselves and future generations from the dangers of radioactive pollution.

Innovative Solutions to Radioactive Pollution Challenges

Radioactive pollution poses significant risks to both the environment and human health. The release of radioactive materials, whether through nuclear accidents or improper waste disposal, can have long-lasting effects on ecosystems and communities. However, scientists and researchers have been tirelessly working towards finding innovative solutions to tackle these challenges and mitigate the impacts of radioactive pollution. This subchapter explores some of the most promising advancements in the field of pollution biology.

One groundbreaking approach is the development of bioremediation strategies that utilize the unique abilities of certain microorganisms to naturally degrade or immobilize radioactive contaminants. These microbes can break down radioactive substances into less harmful forms or trap them in mineral formations, thereby reducing their mobility and potential for spreading. This technique has shown promising results in sites contaminated with uranium, plutonium, and other radioactive elements.

Another area of innovation lies in the development of advanced filtration systems. These systems are designed to efficiently remove radioactive particles from air and water, preventing their spread and minimizing human exposure. Nanotechnology, for example, has offered new possibilities in creating highly effective filters capable of capturing even the smallest radioactive particles. Such advancements have proven invaluable in protecting both the environment and public health.

Moreover, scientists are exploring the potential of phytoremediation, a technique that employs plants to absorb and store radioactive contaminants. Certain plant species have the ability to accumulate radioactive elements in their tissues without being significantly harmed. By cultivating these plants in contaminated areas, we can effectively extract and safely dispose of radioactive materials, thereby purifying the soil and groundwater.

In addition to these biological approaches, technological innovations are also playing a crucial role in addressing radioactive pollution challenges. Robotics and remote sensing technologies are being deployed to access and decontaminate hazardous areas, reducing the risk to human workers. Advanced monitoring systems equipped with sensors and detectors are being developed to provide real-time data on radiation levels and facilitate early warning systems.

While the challenges posed by radioactive pollution are significant, the field of pollution biology is continuously evolving to find innovative solutions. These advancements offer hope for a cleaner and safer future, where the impacts of radioactive pollution can be effectively mitigated. By combining scientific knowledge, technological advancements, and a commitment to environmental stewardship, we can work towards a world where the dangers of radioactive pollution are minimized and our ecosystems and communities are protected.

Chapter 8: Future Perspectives and Concluding Remarks

Advances in Radioactive Pollution Research

In recent years, there have been remarkable advancements in the field of radioactive pollution research, shedding light on the complex nature of this environmental issue. The detrimental effects of radioactive pollution have become a growing concern for both scientists and the general public. As a result, researchers from various disciplines, particularly in the field of pollution biology, have been working tirelessly to uncover the hidden truths behind this silent threat.

One significant breakthrough in radioactive pollution research is the development of sophisticated detection and measurement techniques. Scientists now have access to highly sensitive instruments that can accurately detect even the tiniest traces of radioactive substances in air, water, soil, and living organisms. These cutting-edge technologies have enabled researchers to monitor the levels of radioactivity in different environments, aiding in the assessment of potential risks and the development of effective mitigation strategies.

Furthermore, recent studies have focused on understanding the ecological impact of radioactive pollution. Researchers have conducted extensive field studies, documenting the effects of radiation on various organisms and ecosystems. These studies have highlighted the intricate interplay between radiation levels and biological systems, providing valuable insights into the long-term consequences of exposure to radioactive substances. Moreover, researchers have explored the genetic and physiological mechanisms that allow certain organisms to adapt and survive in radioactive environments, potentially leading to the discovery of new strategies for environmental remediation.

Advances in radioactive pollution research have also contributed to the development of innovative cleanup technologies. Scientists have been exploring novel approaches, such as phytoremediation, which involves using specific plant species to absorb and store radioactive contaminants. This eco-friendly technique has shown promising results in contaminated areas, offering a more sustainable and cost-effective solution compared to traditional cleanup methods.

Furthermore, the integration of big data analytics and modeling techniques has revolutionized radioactive pollution research. Scientists are now able to collect and analyze vast amounts of data, facilitating a deeper understanding of the complex factors influencing the transport and fate of radioactive substances in the environment. This interdisciplinary approach has allowed researchers to make more accurate predictions and develop comprehensive risk assessment models, providing valuable information for policymakers and stakeholders in developing effective strategies for prevention and mitigation.

In conclusion, the field of radioactive pollution research has witnessed significant advancements in recent years. These breakthroughs have not only enhanced our understanding of the detrimental effects of radioactive pollution but also paved the way for more effective monitoring, mitigation, and cleanup strategies. With ongoing research and collaboration, scientists are hopeful that further progress will be made, offering a safer and healthier environment for all.

Emerging Technologies for Radioactive Pollution Detection and Control

In recent years, the world has witnessed a growing concern regarding the impact of radioactive pollution on our environment and health. As advancements in technology continue to shape our lives, innovative solutions are being developed to detect and control radioactive contamination. This subchapter explores some of the emerging technologies that are revolutionizing the field of radioactive pollution detection and control.

One of the most promising technologies in this area is the use of drones equipped with advanced sensors. Drones can be deployed to survey large areas quickly and efficiently, providing real-time data on radioactive hotspots. These drones can detect and map radiation levels, helping experts identify contaminated areas and prioritize cleanup efforts. By utilizing drones, we can minimize human exposure to radiation and enhance the effectiveness of remediation strategies.

Another exciting development is the application of artificial intelligence (AI) in radioactive pollution detection. AI algorithms can process vast amounts of data and identify patterns that may indicate the presence of radioactive materials. This technology can be integrated into existing monitoring systems, enabling early detection of contamination and rapid response. By leveraging AI, we can enhance the accuracy and efficiency of radioactive pollution detection, leading to more effective control measures.

Furthermore, nanotechnology offers immense potential in the field of radioactive pollution control. Nanomaterials can be engineered to selectively absorb or neutralize radioactive particles, effectively reducing their impact on the environment. These nanomaterials can be incorporated into filters, protective clothing, and even building materials, offering passive protection against radiation exposure. The development of nanotechnology-based solutions opens up new avenues for minimizing the risks associated with radioactive pollution.

Lastly, the advent of blockchain technology has the potential to revolutionize the transparency and accountability of radioactive waste management. Blockchain, a decentralized and immutable ledger, can track the entire lifecycle of radioactive materials, from production to disposal. This technology ensures that all transactions and movements of radioactive materials are recorded and accessible to relevant stakeholders, minimizing the risk of illicit activities and unauthorized disposal.

The emergence of these technologies presents new opportunities for combating radioactive pollution and safeguarding our environment. By harnessing the power of drones, AI, nanotechnology, and blockchain, we can enhance our ability to detect, control, and manage radioactive contamination. As we continue to advance in these fields, it is crucial to invest in research and development to unlock the full potential of these technologies and ensure a safer and cleaner future for all.

Call to Action: Promoting Environmental Responsibility

In today's world, the issue of environmental responsibility has become more crucial than ever. The impact of human activities on the environment, especially in terms of pollution, is undeniable. As citizens of this planet, it is our collective responsibility to take action and promote environmental responsibility to safeguard our future and that of future generations.

Pollution biology, a niche field of study, focuses on understanding the effects of pollution on living organisms and the environment as a whole. This subchapter aims to provide insights into the science of radioactive pollution and its implications for all readers, regardless of their background or expertise. By understanding the science behind radioactive pollution, we can gain a deeper appreciation of its consequences and be better equipped to make informed decisions.

The call to action for promoting environmental responsibility begins with education. By raising awareness about the causes and effects of radioactive pollution, we can empower individuals to take necessary steps towards minimizing their own impact on the environment. This subchapter will explore the various sources of radioactive pollution, such as nuclear power plants, industrial activities, and improper disposal of radioactive waste. It will also delve into the potential health risks associated with exposure to radioactive materials.

Additionally, this subchapter will highlight the importance of adopting sustainable practices and technologies to reduce radioactive pollution. From renewable energy sources like solar and wind power to proper waste management systems, there are numerous ways individuals and communities can contribute to a cleaner and safer environment. By emphasizing the benefits of these practices, we hope to inspire readers to take action and make environmentally responsible choices.

Furthermore, the subchapter will discuss the role of government regulations and policies in mitigating radioactive pollution. It will shed light on the need for stricter environmental standards and the importance of holding industries accountable for their actions. By advocating for stronger regulations and supporting organizations working towards environmental protection, we can collectively create a healthier and more sustainable future.

In conclusion, promoting environmental responsibility is a call to action that concerns every one of us. Through education, adopting sustainable practices, and advocating for change, we can make a significant difference in reducing radioactive pollution and safeguarding our environment. This subchapter aims to provide valuable insights into the science of radioactive pollution, empowering readers from all backgrounds to take meaningful steps towards a more environmentally responsible future.

Final Thoughts on the Science of Radioactive Pollution

As we conclude this journey into the science of radioactive pollution, it is essential to reflect on the key insights gained and their implications for our understanding of this pressing issue. Throughout this book, we have explored the intricate relationship between radioactive substances and the environment, shedding light on the potential dangers they pose to human health and the delicate balance of our ecosystems.

One of the main takeaways from our exploration is the need for a holistic approach to tackling radioactive pollution. This issue transcends boundaries and affects not only the realms of pollution biology but also public health, policy-making, and environmental management. It is a problem that demands collaboration and cooperation between scientists, policymakers, and the general public.

The science of radioactive pollution has revealed the long-lasting impacts of nuclear accidents, such as Chernobyl and Fukushima, on both the environment and human health. These incidents serve as stark reminders of the devastating consequences of mishandling radioactive materials. They emphasize the importance of strict regulations, robust safety measures, and efficient emergency response systems to prevent and mitigate future disasters.

Furthermore, our exploration has shed light on the often-overlooked sources of radioactive pollution, such as industrial processes, mining activities, and medical applications. These sources contribute significantly to the overall radioactive contamination of our environment and warrant further attention and regulation.

In the face of this complex issue, it is crucial for each one of us to take responsibility and contribute to the solution. By making informed choices in our daily lives, such as opting for renewable energy sources, minimizing exposure to radiation, and supporting sustainable waste management practices, we can all play a role in reducing the impact of radioactive pollution.

Ultimately, this book aims to empower the general reader with knowledge and understanding of the science behind radioactive pollution. By demystifying the subject, we hope to inspire individuals to engage in meaningful discussions, ask critical questions, and demand responsible actions from governments, industries, and themselves.

The science of radioactive pollution is far from complete. As new research emerges and technologies advance, our understanding of this issue will continue to evolve. It is our collective responsibility to stay informed, remain vigilant, and work towards a future where the harmful effects of radioactive pollution are minimized, and our planet is protected for future generations.

In conclusion, the science of radioactive pollution is a call to action for every one of us. By embracing our roles as informed citizens, responsible consumers, and advocates for change, we can make a significant difference in mitigating the consequences of this pressing environmental challenge. Let us all strive to build a future where the harmful effects of radioactive pollution are relegated to the pages of history books, and our planet thrives in a radiation-free environment.

78